YOUR KNOWLEDGE HAS VALUE

- We will publish your bachelor's and master's thesis, essays and papers

- Your own eBook and book - sold worldwide in all relevant shops

- Earn money with each sale

Upload your text at www.GRIN.com and publish for free

Oscar Webb

Hurricane Preparedness Planning for a Hospital

GRIN Publishing

Bibliographic information published by the German National Library:

The German National Library lists this publication in the National Bibliography;
detailed bibliographic data are available on the Internet at http://dnb.dnb.de .

This book is copyright material and must not be copied, reproduced, transferred,
distributed, leased, licensed or publicly performed or used in any way except as
specifically permitted in writing by the publishers, as allowed under the terms and
conditions under which it was purchased or as strictly permitted by applicable
copyright law. Any unauthorized distribution or use of this text may be a direct
infringement of the author s and publisher s rights and those responsible may be
liable in law accordingly.

Imprint:

Copyright © 2013 GRIN Verlag GmbH
Print and binding: Books on Demand GmbH, Norderstedt Germany
ISBN: 978-3-656-97587-8

This book at GRIN:

http://www.grin.com/en/e-book/301299/hurricane-preparedness-planning-for-a-
hospital

GRIN - Your knowledge has value

Since its foundation in 1998, GRIN has specialized in publishing academic texts by students, college teachers and other academics as e-book and printed book. The website www.grin.com is an ideal platform for presenting term papers, final papers, scientific essays, dissertations and specialist books.

Visit us on the internet:

http://www.grin.com/

http://www.facebook.com/grincom

http://www.twitter.com/grin_com

Hurricane Preparedness Planning

Hurricane Preparedness Planning

Oscar Webb

Saint Leo University

Abstract

Implementation of a hurricane preparedness and response plan requires the coordinated efforts of multiple agencies from throughout the region. Utilization of effective team support produced very good results when applied with strategic management efforts built on an application of solid ethical principles. Understanding the importance of strong communication throughout the process allowed the team to produce superior results and build upon previous successes. The team was also cognizant that the experiences of the community and those that had gone through the process previously were important factors in allowing the current process to proceed in a positive fashion.

Hurricane Preparedness Planning

Disaster preparedness is by no means an exact science. Despite this fact, there are ways to

eliminate errors and provide an organization with the highest chance to reduce negative

outcomes. This is achieved through experience and repetitive training designed to mimic

potential scenarios that may be encountered. Analyzing previous successes and failures of the

organization and other entities can provide valuable insight into what strategies will lead to the

desired outcomes for the planning committee. Failing to consider the merits of past performance

is a surefire way to mismanage a potential impending disaster. Even a successful previous plan

can yield data about corrections and improvements when analyzed thoroughly.

The Saint Leo University Hospital is faced with the prospect of preparing a disaster plan

for the cycle of hurricane season. A successful disaster and response plan will require the input

of many professionals from various departments within the facility. With this in mind, one of the

first tasks is identifying the individuals that should be included in the planning process. The

American Society for Healthcare Engineering has outlined the required members of a disaster

planning committee as: (medical staff, risk manager, nursing staff, emergency department,

security, communications, public relations, medical records, engineering, laboratory, radiology,

and respiratory therapy (American Society for Healthcare Engineering Webpage. (2014).

Retrieved March 19, 2014 from

http://www.ashe.org/advocacy/organizations/TJC/ec/emergency/hospdisasterprepare.html).

Many concepts are useful in the construction of the planning team and subsequent development of the eventual disaster and recovery plan. One of the touted models is "complex systems thinking" (Nebraska Symposium on Motivation, Shuart, J. W., Spaulding, W. D., & Poland, J. S. (2007). Modeling complex systems. Lincoln [Neb.: University of Nebraska Press.). The complex systems approach recognizes the different levels that the various components of the recovery effort operate at, and seeks to interlink these critical elements in a way that elevates their combined effectiveness above what they could achieve on an individual level (Burns, L., Bradley, E., & Weiner, B. (2012). *Shortell & Kaluzny's Health Care Management: Organization Design and Behavior* (6th Ed.). Clifton Park, NJ: Delmar).   This aggregate effect is possible because the complex systems approach also allows the group to be more effective as the group is more receptive to change because the diversity is a strength that fosters adaptation (Advances in complex systems. (1998). Singapore: World Scientific Pub.). In a setting such as disaster planning, it is important to be able to entertain multiple points of consideration before becoming settled on a particular plan of action. Being able to give every member of a group due diligence to both express their ideas and consider the offerings of others is important for a successful planning outcome.

In setting the tone for crafting the execution of the disaster plan, the theories and roles of leadership will play a significant role. The roles of the personnel involved in the planning process and the strategy employed by management are two of the most important factors to be considered. Although the hospital has decided to utilize a complex systems approach to creating the disaster plan, it is still a good idea to outline how each group of individuals will contribute to the effort. Making sure everyone is on the same page regarding roles is an important step prior to the start of the evolution. This will eliminate wasting time explaining job functions to team members, or having members duplicating efforts. In order to expand the ability of the healthcare facility to respond to a pending disaster, it may be prudent to identify inter-organizational relationships (IOR's) that can be beneficial for securing resources and manpower assistance (Martens, R., Heene, A., & Sanchez, R. (2008). Competence building and leveraging in interorganizational relations. Amsterdam: Elsevier JAI.) . Because an event like a hurricane has the potential to impact residents of the entire community, the staff should not limit the participating organizations to just those with a previous history of providing assistance. Based on the identified needs related to expected damage, the staff should contact those agencies that may be able to provide the type of resources that will allow the community to mitigate the aftermath with minimal difficulty.

As the staff begins to develop the strategy for the crafting of the hurricane plan, it is important to make sure the objectives are clearly stated. This is important for the members of the St Leo Hospital staff and the other agencies being recruited to assist in the process. Establishing unambiguous objectives gives a good starting point for how the organization wants to approach allocation of resources and execution of tasking. Management should designate representatives from each department to ensure this function of the process is properly addressed before other areas are started. More detailed strategies cannot be developed until the roles are firmly established and each center has a clear understanding of who is in control of what functions.

As the organization further develops its plan for hurricane preparation, it must consider additional factors related to organizational design and employee motivation. For the proper utilization of organization design, there must be a concerted effort to identify the tasks important to the process and segment them accordingly. The first level will occur by dividing work between those jobs that can be categorized as either professional or nonprofessional in nature. The professional staff jobs will consist of the physicians, nurses, administrators, and IT personnel. The nonprofessional staff will be the workers that perform the support positions around the facility.

During a crisis such as a hurricane, it is important to be efficient in delivering services. With this in mind, it is a good practice to utilize an integrated delivery system, if one is not already in place. The benefit of such a system is that it will allow management to ensure critical resources are delivered across the entire region that the system is responsible for responding to. This allows the delivery system to coordinate with the various healthcare facilities to avoid mismanagement of supplies and to better work the time element that is so crucial in the wake and aftermath of a hurricane. The need to be proactive with partnering agencies must be emphasized in order to allow sufficient time for all organizations to work with both internal and external entities in coordinating resource planning.

In the process of coordinating, there must be an understanding of the ability level of the personnel involved at each level. Management must be certain they are staffing personnel at the appropriate skill levels. During the middle of a hurricane is the wrong time to discover a person does not know how to lead an evolution. The person's slow reaction time could jeopardize lives and the safety of the public at large. In addition to jeopardizing lives, individuals must understand how to manage resources in a crisis versus how they manage resources during normal operating conditions. During the training portion of the time prior to the actual implementation of the final plan, the management staff should focus on using the process of relational coordination as a way of getting individuals that will be working together on projects to gain a better understanding of each other's role within their respective departments. As the agencies allocate members of their teams to fill positions to assist in the work of preparing for disaster mitigation, the team members that have previous working experience will understand the strengths and weaknesses of their peers. The team members that are being paired together for the first time will need to establish an understanding of what their contemporary's abilities are in order for the new team to operate at its most efficient level. This use of relational coordination allows the members to identify what these areas of strength and weakness are, and therefore put the entire process in a better position to achieve a positive outcome in the event of a real crisis (Burns, L., Bradley, E., & Weiner, B. (2012). Shortell & *Kaluzny's Health Care Management: Organization Design and Behavior* (6th Ed.). Clifton Park, NJ: Delmar).

During the planning process, management will be tasked with motivating their staff to perform at levels greater than normal. When an actual incident occurs, the staff will have to continue to be motivated to perform their duties under conditions that will require them to make sacrifices within themselves to keep going at times. As a manager, the needs and motivations of the workers must be understood in under to keep them being productive under the most undesirable conditions. Motivation cannot be used as a tool to create productivity. Management must be aware of the talents of the individual workers and constantly train and put these workers in a position to succeed. The aggregate of the skills of these workers means nothing if it isn't coordinated properly and effectively, allowing the totality of their abilities to have a meaningful impact on the project of importance. There are ways to effectively motivate employees, such as allowing them to participate in the goal setting process. When management isolates employees from the goal setting process and then hands them a set of expectations, it can lead to resentment. By incorporating employees in the goal setting process, it also allows for greater accountability if the metrics are not met. When employees are handed goals they were not part of creating, they can easily dismiss them when the goals are not met by saying the management was unreasonable when they devised them. Another element of motivation for professional staff seems to be the level of autonomy afforded to them (Dibachi, F., & Dibachi, R. (2003). Just Add Management: Seven Steps to Creating a Productive Workplace and Motivating Your Employees in Challenging Times. New York: McGraw-Hill.).

A challenge for all managers is to find the right combination among their staff to develop an atmosphere where "effective team support" becomes a normal practice. All organizations are staffed with individuals with varying levels of ability; even the highest skilled personnel require the coordinated efforts of those around them in order to allow their efforts to be maximized. Management has the responsibility of putting these teams together to overcome obstacles and allow the planning and relief efforts to operate at peak efficiency. External factors play a major role in organizations being able to effectively deal with high pressure situations. The environment that an agency has to work in plays a major role in the outcome of the combined efforts of all agencies involved in the process. The internal and external communication is equally important in the developmental process of a healthcare facility during a disaster preparedness effort. Managers must be proactive in ensuring that all members of the team are carrying their weight and that no member of the team can be designated as a "free rider". When all team members are actively participating it gives the team a greater sense of camaraderie. This is the latest sense of teamwork can carry the team forward when they have to deal with difficult times such as long hours and difficult situations.

In order to effectively deal with a hurricane if it affects the immediate community, the hospital must be able to communicate not only with the staff and the agencies that have partnered to assist in the aftermath, but also with the public at large. This is an important distinction that cannot be overlooked in the planning process. The assistance of the public can be very beneficial to minimizing the effects that such a damaging storm can have to the immediate area. Without the public support, the facility can become overwhelmed in a very short amount of time. Communicating as much information as possible related to prevention efforts can be a key to allowing people to buffer themselves against damage. There is no reason for the disaster response teams to withhold valuable information from the very people whose lives could be saved with the knowledge. To assist with the sharing of information, there should be a central location established where all official information is disseminated from. Other data can be filtered into the location, but nothing gets passed out to media or the public unless it comes from the central location. This is the best way to control the flow of information; it also avoids misinformation being given out and allows the public to stay informed on the current status of preparations. The disaster preparation team will likely want to establish some type of toll-free number for the public to use to access information related to the disaster preparation effort (Landesman,L. American Public Health Association Website: Public Health Management of Disasters: The Pocket Guide.(2014). Retrieved April 20, 2014 from http://www.apha.org/NR/rdonlyres/ECDFA2EC-49A0-4D7A-B257-1488E671DE5B/0/APHA_DisasterBook_2.pdf).

While managing the communication efforts of the disaster plan many agencies will have a role and therefore proper feedback is more crucial than ever. Every individual involved in the process should ensure the message they receive and deliver is understood by the receiver each time they communicate. There are many barriers to proper communication however these barriers must be overcome through proper and effective communication methods. Some members will have a greater amount of knowledge than others and feel as though other staff members should also have an equal amount of knowledge. They may not share information that the group is not privy to. This lack of sharing may not be intentional however it may still have a very damaging effect on the group's ability maintain a cohesiveness moving forward. When the team members on a home on information to grow and to outside staff members they must always remember to evaluate is receiving information and level of knowledge of the receiver. Team members must work to minimize distractions that will take away way from the performance of their duties. The most effective teams understand that cohesiveness can only be achieved when they avoid allowing outside interference affect their ability to communicate with one another.

Power will undoubtedly play a role in the creation of the response plan for the hospital. Understanding the importance of power and how to utilize it without allowing it to turn into something negative is important for the entire project. Power is necessary for allocation of resources, and also in order to get workers to respond to requests for action in a timely manner. This power must come from an appropriate source that gives the individual effective authority to create the type of change that will help the team grow and realize all of their strengths. The disaster preparedness effort cannot be led by individuals that lack effective use of power, or worse, are powerless. There have to be the right individuals put in key positions in order for projects to move forward in a timely manner. Even though this need for power is real, it also must be constrained at times. The personnel given the authority must know when to push certain buttons to get results. They must avoid alienating the individuals that are their greatest source of support by exhibiting a lack of control or respect for their power. An important term is "power stratification" (Burns, L., Bradley, E., & Weiner, B. (2012). *Shortell and Kaluzny's Health Care Management: Organization Design and Behavior.* (6th ed.). Clifton Park, NY: Delmar Centage). This term refers to the power that various stakeholders have in a healthcare organization. Stakeholders have many ways to impact the disaster planning and preparation efforts. Many stakeholders choose critical operating times to demonstrate the extent of their power within the organization. The hurricane planning process is a good time for all stakeholders to participate by contributing their skills to the process. Accumulating power is not just for times of operational strength, but also times when the organization needs leaders to step up and provide direction to the rest of the staff.

In addition to power, politics also plays a major role in forming a committee based hurricane preparedness plan. Political figures may have good intentions or attempt to obstruct efforts by individuals they oppose for unknown reasons. One of the reasons for issues between individuals in authority is a conflict of power. If the use of political power is not demonstrated to be fair and equitable, those in partnership may lose trust and the alliance will suffer in the process. When constructing an alliance for the hurricane response team, it is important that although it will be impossible to avoid keeping politics out of the equation, politics should not become the factor that slows the project down or makes everyone's job harder. Political infighting should be suppressed by strong and competent leadership in order to allow the project to meet its stated goals and take advantage of all of the high level people working in unison for the common cause of preparing to minimize the effects of the hurricane season.

When conflicts do occur, the managers and supervisors must utilize effective tactics to eliminate the conflict so that teams can resume the business of their particular segment. Managers must be able to identify the source of the conflict in order to effectively deal with disagreements. Understanding the mindset of the individuals involved in the conflict can be difficult, but it also can be one of the elements of relationship conflict that a manager needs to be able to interpret to move a team through to get a project back on track. When there are so many different personalities and age disparities brought together, there are bound to be times of disagreement. Therefore the issue is not whether there will be conflict, but how to handle these disagreements when they happen. In a setting of healthcare planning, emotion can be a dangerous element; it is therefore even more important to swiftly identify the source of a conflict between individuals and suppress the conflict without choosing sides, or allowing either person to feel a sense of heightened anger. Any lingering resentment in a group setting will be felt by the rest of the group and will affect the productivity of the rest of the members.

Aside from conflict, the medical team will want to exercise influence over members of other agencies in order to get better results for a final plan. The other agencies will also attempt to use their staff members to alter the course of the proceedings, thus makings the challenge of making change more complicated. Even in a team setting, personnel will form alliances based on the interests of their particular stakeholders. It is important to show some measure of compromise without totally abandoning one's original position. All team members must be willing to give a little in the process in order to get back some of the things they feel are important to their constituents.

While progressing with efforts to develop the hurricane disaster plan for the hospital, it must be understood that the dynamics of a healthcare facility are very different than other types of organizations. Healthcare organizations by nature are very complex, and where mistakes in other institutions may be uncommon and cause for alarm, this is not the case in this industry. That is not to say that mistakes are acceptable or part of the routine, just that healthcare is a constantly evolving field where professionals can learn from mistakes and apply the changes on a wide scale for the greater good. Often when a process doesn't work, it eventually leads to a better process. With so many new issues arising in healthcare each day, it's impossible to predict how each day is going to go. No facility sees the same number of patients for the same set of conditions. The frequency and seriousness changes each day, therefore making scheduling very difficult; this is just one of the difficulties a medical facility has to deal with. The uniqueness of the complex systems of healthcare organizations is the fact that they exist in a state that cannot be described as being totally orderly or totally chaotic. Some managers may enter this world and feel an inclination to bring massive change to bring the organization closer to a state more associated with traditional order. This line of thinking typically is not successful in healthcare due to the nature of the work being so unpredictable. Being able to have an understanding of what all of the various parts are doing, and how they are acting in coordination with one another is probably the most the manager can hope to achieve in the organization. During the hurricane disaster planning effort, the central planners must be aware that there will be a great number of moving parts once the central command goes live. Being in total control of element is simply not going to happen.

The training process is an important factor in the success of the hurricane preparation plan. Training will be conducted continuously by the individual agencies for their staff, and also on a larger scale for volunteers and the public that desires assistance with pre-storm help. The information that will go into the training from the St Leo University Hospital staff is based on a process of double-loop learning (Burns, L., Bradley, E., & Weiner, B. (2012). Shortell and *Kaluzny's Health Care Management: Organization Design and Behavior. (6*th *ed.).* Clifton Park, NY: Delmar Centage). This process has allowed the facility to go in depth in evaluating the policies and procedures that have been in place for many years, and how these policies have impacted events previously. The results of the evaluation are available for review by the public at-large. The analysis shows that organizational policy has been vital in creating opportunities for successful outcomes in some instances, while producing less favorable conditions at other times. Some of these issues include the way medical providers manage the record systems of patients, including the time frame for making updates. Some patients have transferred or requested referrals without the required information being available to the new physician. These and other issues have prompted calls for additional training in correct procedures leading up to the implementation of the hurricane preparedness plan.

The best coordination of efforts requires the use of the best innovation methods available. The best innovation means the best thoughts, ideas or technology. All innovation begins with a thought or idea before it can be transferred into an actual product. The planning team is open to the use of any relevant technology that can benefit the relief efforts. This applies to all facets including medical care, delivery of supplies, resource allocation, communication, IT support, or storm related prevention. One of the issues with innovation and the development of ideas is in a group setting, when an idea is presented, that idea can become altered or expanded by the input of do many other voices. Although the original idea may have been highly innovative and beneficial, it may become lost or so different that it no longer has the same impact as it originally intended. This expansion of the original idea may happen as a result of the group's enthusiasm resulting from the original idea, and not as a desire to dismiss it. The person presenting the idea must return to their idea, stressing the merits of it, and getting the team to focus on the task of crafting the best plan possible, not just creating more ideas (Di Tommaso, M. R., & Schweitzer, S. O. (2005). Health Policy and High-tech Industrial Development : Learning Form Innovation in the Health Industry. Cheltenham, UK: Edward Elgar.).

Strategic management is an important concept in the development of the hurricane preparedness plan. The use of strategic planning is necessary to ensure that the goals on the organization are intertwined in the framework of the planning document. Without strategic planning, this cannot be determined or assured. It is not wise for the manager to go back and determine if the information was included after the finished document has been delivered. At that point it's too late to make the changes, and the organization will not have their goals addressed by the structure of the planning actions. Putting a strategic plan in place occurs over a long period of time utilizing a serious of carefully thought out plans vetted on many levels of the organization. The strategic plan will include an evaluation of events that have happened in the past, with revisions to allow for better results in the future. Managers, and those assigned to carry out the planning process, must be aware that the strategic planning process is very fluid. What this means is that it must not be allowed to become stagnant, it must be constantly evaluated while the hurricane preparedness plan is being crafted. Although the organization has a plan in place with a set of goals they hope to achieve, they must mesh these goals with the preparedness plan and work to ensure they allow the goals to be included without obstructing the overall process.

Inter-organizational collaboration is an important ingredient in how successful the project will turn out, and also how efficient the process will be overall. There will be a number of barriers to achieving an optimal level of organizational collaboration. Some of these are similar to what will be found with an alliance, such as mistrust or differences in agendas. It may be difficult to overcome these differences initially, but the team leaders must be proactive in taking control of the group and aligning all members with the goals of the group as a whole. Although the individual members represent various interests, the team leaders must get them to place the good of the team above those of their individual organizations. That is not to say that the final plan will not include elements that support the goals of those various organizations. In many instances the agency goals are similar to those of the group at-large, making compromise very easy (Linden, R. (2002). Working Across Boundaries : Making Collaboration Work in Government and Nonprofit Organizations. San Francisco, Calif: Jossey-Bass.). Getting to a point where people can see that their values are the same as others in the group is often a turning point for creating compromise and actually being productive in a collaborative setting.

The values, vision, and mission of the collective organizations will shape how the hurricane preparedness plan is created. In addition to the values of the organizations, the community must also be taken into account in the planning process. Understanding the needs and limitations of the population that is being served is a critical part of the planning process. Also knowing the way they are shaped by their values is also important when planning for things like delivery of care and resources. Depending on the demographics of the community, the needs and capabilities may be increasingly dependent on the first responders. In other places, the residents may have a greater capacity to provide assistance not just to themselves, but to those around them. The geographic area also plays a role in the delivery of care and resources. The mission of the St Leo Hospital is to ensure that all residents are cared for in a timely, safe, professional manner. In order to do this, all members of the planning process must be fully engaged and share in this vision for providing quality care despite the difficult conditions. The values of each agency must be exhibited in the planning process. In seeking to inject their value in the process, each agency must be careful not to take a position of being selfish, or trying to get something out of it. In a disaster preparedness planning team, there must be a sense of mission on a level that extends beyond just the member's individual organization. In fact, an organization with a mission to provide the most good to its patients should want to fully invest in the disaster response effort to give them as much of a voice as possible.

Regulation of public health processes has been going on for many years, and will continue to do so, although with a great deal of scrutiny by those on both sides of the issue. When planning hurricane relief efforts, it is imperative to stay abreast of all regulatory issues related to delivery of care, payment processing, environmental concerns, safety issues, and others community specific issues. There are a number of different programs that the federal government is responsible for administering and paying out. Some of the most widely used include Medicare and Medicaid. The widespread use of these programs makes it critical that the hospital have a plan in place to handle the increased volume of patients that may result from a hurricane in the areas. The staff needs to know the procedures associated with these programs in order to avoid problems with compliance down the line. In addition to challenges with processing patients and payments, the staff must be on prepared for the increased volume of prescription requests that will be submitted as a result of injuries and infections. Patients have a right to receive care, even during a time of crisis. Although there may be a reasonable expectation that many of these services will come at a slower rate, patients still expect to get treated when they arrive at a facility. Regulators will be contacted if the facility is unable to handle the volume of patients, even under these difficult conditions. Also, although the facility is experiencing a disaster relief effort, the professional staff must take care to process Medicare and Medicaid payments based on required standards. Anything less than this can find a facility in a lot of trouble when the auditors arrive.

One of the standards of the medical field is the Health Insurance Portability and Accountability Act (HIPAA) law. This ensures that patient information is kept private and medical facilities maintain minimum levels of information security to ensure that information does not get disclosed to unintended parties. This law remains intact at all times, even during a natural disaster and its aftermath. The law provides additional protections that must also continue to be adhered to doing the recovery stage. It is important to uphold these standards throughout the process as they will be evaluated as part of the facility's ability to operate according to professional standards (National Academies Press, (. (U.S.), & Institute of Medicine, (. (U.S.). (2011). For the Public's Health : Revitalizing Law and Policy to Meet New Challenges. Washington, D.C.: National Academies Press.).

The effective use of health information systems will allow the implementation of strategic goals to be done in a more efficient manner. Being that the project will be bringing in organizations from various entities, it is important that any health information system be implemented on a regional scale. The planning for such an endeavor will fall into the hands of the management team and the IT staff of the organizations. The system should at the very least consist of an integrated communication model where the various agencies can deliver information in real-time. Being able to communicate with one another in a timely manner is important for operations to function and flow without interruption. It also allows everyone to be on the same page and receive information at the same time. When dealing with a hurricane relief scenario, providers will be dealing with not only time critical situations, but a need for patient information that is accurate. This is where health information systems are most important. The financial barriers to establishing an electronic health record system is well documented. In the wake of a natural disaster, the collective efforts of all the involved agencies to request funding to implement a system on a regional level could be the difference in getting an EHR implemented.

Effective project management will be needed if the planning team decides to pursue an information system implementation strategy. Selecting the correct individual to manage the project is part of the process. It is also important to detail what the intentions and objectives of the project are. The rest of the implementation team will follow the lead of the management, and therefore, an effective project manager can direct the activities based on the goals that have been set for the direction of the information systems expansion. Also, assigning the correct personnel to manage the finances of the project is important so that the project doesn't end up going over budget (Blobel, B. (2002). Analysis, Design and Implementation of Secure and Interoperable Distributed Health Information Systems. Amsterdam: IOS.) .

The community has a shared responsibility in the hurricane preparedness planning process. Because so many people are potentially affected, the personal responsibility of every citizen in the affected area must be highlighted. Citizens must take advance action in doing the things that they can do in advance of the approaching storm. These things include, but are not limited to: filling prescriptions for themselves and their loved ones, checking on the needs of hospitalized family members and obtaining supplies for home; waiting until the storm has arrived is too late. Also, expecting the hospital or other agencies to provide critical resources may be a bad idea. This is especially true in the event of a storm that produces long term effects for the area. If this happens, whatever resources agencies have will be expended very quickly to those with the greatest need. It is therefore imperative that the public exercise as muse responsibility for themselves as possible (Burns, L., Bradley, E., & Weiner, B. (2012). Shortell *and Kaluzny's Health Care Management: Organization Design and Behavior. (6*[th] *ed.).* Clifton Park, NY: Delmar Centage).

The sharing of information will be frequent during the recovery process. Every professional is aware of the HIPAA law, but must be still reminded of the greater standard of the "need to know". This means that just because someone is a medical provider doesn't mean that medical information should be released to them. An individual's personal medical history should be safeguarded and released only when the information has a reason to be released to an individual. These principles cannot be sacrificed because of the conditions that the staff is working under. As professionals, team members must work with each other to help each other during difficult working conditions, not allowing a peer to make serious discrepancies when they can be avoided.

Every action that drives what decision is made is dictated by the ethical principles that healthcare professionals are shaped by. Allowing decisions and actions to be based on ethics ensures that they will be in line with the values of both the individual and the profession. This will create good policy that can withstand scrutiny and can be crafted to protect the interests of the patients that are in the need of the critical care. All providers must be honest and faithful in their dealings with each other and the community. By adhering to this principle the providers create a level of trust with the community that will allow them to do their job at a more effective level. The staff must also respect the autonomy of the patient and recognize that there may be times when a patient may refuse treatment. This is within the rights of the patient as long as there is no question about the competence of the patient.

Medical innovation and patient flow occurs even outside of U.S. borders. During the process, the team may need to inquire with outside agencies for additional assistance regarding information support. These outside agencies can also provide outsourcing support to assist with support services. They may even be able to send professional personnel to assist with patient care. Some of these services may be financial services related to payment processing, resource allocation, or IT support. Utilizing the experience of those that have gone through the process before, those that are enlisted to assist will not have to develop strategies with no background knowledge. Throughout the process, there will be personnel that have experience in various stages of the planning process that they will be able to share with their teammates. Having this ability to share experience will make the collaboration process go much smoother. Jurisdiction must be considered when constructing any part of the response plan, as certain geographic areas will have specific responsibilities, however, there must still be certain strategic, tactical, and organizational strategies put in place and successfully coordinated on all levels. The St Leo Hospital must carefully evaluate the capabilities of the support organization and determine exactly how to utilize their strengths to best support the goals of the hurricane relief effort. The community planning team should be included in this effort to allow them input on how to

coordinate their needs into the services the outside agency is offering (FEMA Website (2014)

Developing and Maintaining Emergency Operation Plans. Retrieved April 20, 2014 from

http://www.fema.gov/media-library-data/20130726-1828-25045-

0014/cpg_101_comprehensive_preparedness_guide_developing_and_maintaining_emergency_o

perations_plans_2010.pdf).

References

Advances in complex systems. (1998). Singapore: World Scientific Pub.

American Society for Healthcare Engineering Webpage. (2014). Retrieved March 19, 2014 from http://www.ashe.org/advocacy/organizations/TJC/ec/emergency/hospdisasterprepare.html.

Blobel, B. (2002). Analysis, Design and Implementation of Secure and Interoperable Distributed Health Information Systems. Amsterdam: IOS.

Burns, L., Bradley, E., & Weiner, B.(2012). *Shortell & Kaluzny's Health Care Management: Organization Design and Behavior* (6th Ed.). Clifton Park, NJ: Delmar.

Dibachi, F., & Dibachi, R. (2003). Just Add Management : Seven Steps to Creating a Productive Workplace and Motivating Your Employees in Challenging Times. New York: McGraw-Hill.

Di Tommaso, M. R., & Schweitzer, S. O. (2005). Health Policy and High-tech Industrial Development : Learning Form Innovation in the Health Industry. Cheltenham, UK: Edward Elgar.

FEMA Website (2014) Developing and Maintaining Emergency Operation Plans. Retrieved April 20, 2014 from http://www.fema.gov/media-library-data/20130726-1828-25045-0014/cpg_101_comprehensive_preparedness_guide_developing_and_maintaining_emergency_operations_plans_2010.pdf.

Landesman,L. American Public Health Association Website: Public Health Management of Disasters: The Pocket Guide.(2014). Retrieved April 20, 2014 from http://www.apha.org/NR/rdonlyres/ECDFA2EC-49A0-4D7A-B257-1488E671DE5B/0/APHA_DisasterBook_2.pdf.

Linden, R. (2002). Working Across Boundaries : Making Collaboration Work in Government and Nonprofit Organizations. San Francisco, Calif: Jossey-Bass.

National Academies Press, (. (U.S.), & Institute of Medicine, (. (U.S.). (2011). For the Public's Health : Revitalizing Law and Policy to Meet New Challenges. Washington, D.C.: National Academies Press.

Martens, R., Heene, A., & Sanchez, R. (2008). Competence building and leveraging in interorganizational relations. Amsterdam: Elsevier JAI.

Nebraska Symposium on Motivation, Shuart, J. W., Spaulding, W. D., & Poland, J. S. (2007). Modeling complex systems. Lincoln [Neb.: University of Nebraska Press.